Ininatig's Gift of
SUGAR

Ininatig's Gift of SUGAR

Traditional Native Sugarmaking

WE ARE STILL HERE
NATIVE AMERICANS TODAY

Laura Waterman Wittstock
Photographs by Dale Kakkak
With a Foreword by Michael Dorris

Lerner Publications Company ○ Minneapolis

Series Editor: Gordon Regguinti
Series Consultant: Juanita G. Corbine Espinosa
Illustrations by Carly Bordeau

This book is available in two editions:
Library binding by Lerner Publications Company
Soft cover by First Avenue Editions
241 First Avenue North
Minneapolis, MN 55401

ISBN: 0-8225-2653-0 (lib. bdg.)
ISBN: 0-8225-9642-3 (pbk.)

LIBRARY OF CONGRESS CATALOGING-IN-PUBLICATION DATA

Wittstock, Laura Waterman.
  Ininatig's gift of sugar : traditional native sugarmaking / Laura
Waterman Wittstock ; photographs by Dale Kakkak ; with a foreword by
Michael Dorris.
     p.   cm. — (We are still here)
  Includes bibliographical references.
  Summary: Describes how Indians have relied on the sugar maple tree
for food and tells how an Anishinabe Indian in Minnesota continues
his people's traditions by teaching students to tap the trees and
make maple sugar.
  ISBN 0-8225-2653-0
  1. Ojibwa Indians—Social life and customs—Juvenile literature.
2. Sugar maple—Minnesota—Tapping—Juvenile literature. 3. Maple
syrup—Minnesota—Juvenile literature. 4. Maple sugar—Minnesota—
Juvenile literature. [1. Ojibwa Indians—Social life and customs.
2. Indians of North America—Minnesota—Social life and customs.
3. Maple sugar. 4. Maple syrup.] I. Kakkak, Dale, ill.
II. Title III. Series
E99.C6W68   1993
338.1'7364'09776—dc20   92-37980
                                                                CIP
                                                                 AC

Manufactured in the United States of America

1 2 3 4 5 6 – P/JR – 98 97 96 95 94 93

Dedicated in thanks, for sharing his wisdom, to Walter Gahgoonse White, his forebears, and his children.

Foreword

by Michael Dorris

How do we get to be who we are? What are the ingredients that shape our values, customs, language, and tastes, that bond us into a unit different from any other? On a large scale, what makes the Swedes Swedish or the Japanese Japanese?

These questions become even more subtle and interesting when they're addressed to distinct and enduring traditional cultures coexisting within the boundaries of a large and complex society. Certainly Americans visiting abroad have no trouble recognizing their fellow countrymen and women, be they black or white, descended from Mexican or Polish ancestors, rich or poor. As a people, we have much in common, a great deal that we more or less share: a recent history, a language, a common denominator of popular music, entertainment, and politics.

But, if we are fortunate, we also belong to a small, more particular community, defined by ethnicity or kinship, belief system or geography. It is in this intimate circle that we are most "ourselves," where our jokes are best appreciated, our special dishes most enjoyed. These are the people to whom we go first when we need comfort or empathy, for they speak our own brand of cultural shorthand, and always know the correct things to say, the proper things to do.

Ininatig's Gift of Sugar provides an insider's view into just such a world, that of the contemporary Anishinabe people of Minnesota. If we are ourselves Anishinabe, we will probably nod often while reading these pages, affirming the familiar, approving that these people keep alive and pass on the "right" way to make maple sugar. If we belong to another tribe, we will follow this special journey of initiation and education with interest, gaining respect for a way of doing things that's rich and rewarding.

This is a book about people who are neither exotic nor unusual. If you encountered them at a shopping mall or at a movie theater they might seem at first glance like anyone else, American as apple pie. *Ininatig's Gift of Sugar* does not dispute this picture, but it does expand it.

Michael Dorris is the author of *A Yellow Raft in Blue Water, The Broken Cord,* and, with Louise Erdrich, *The Crown of Columbus.* His first book for children is *Morning Girl.*

*I*n the Ojibway language of the Anishinabe people, the story is told of *Ininatig*, "the man tree." As a reminder of the importance of lifesaving food in the harsh winters of the northern woods, the Anishinabe, whose name means "the original people," tell the story of maple sugar.

*I*t was the end of a long, cold winter, and a family was starving. The hunting had not been good that year, and all the stored food had been eaten. There were no grocery stores then. As the family looked out at the lake near their camp, they noticed that the ice was changing color, from white to black. This meant that the ice was thin and would break apart in a few weeks. Spring was coming. They would find food then, if only they could stay alive that long.

Behind them, the family heard the trees creaking in the wind. They heard a woodpecker tapping on a tree, looking for insects. Above the noise of the trees, the father thought he heard someone speak. He turned but saw no one. He thought that hunger must be making him hear things.

All of a sudden the mother heard a noise. She asked her husband, "Did you say something?"

"No, I didn't," he said.

They both turned when they heard someone say, "I will teach you a way to make food so that you will never have to starve." The whole family was surprised and frightened. Trees don't talk to human beings! Yet, it was true. They had all heard it. Ininatig—the man tree—had spoken.

He told the family to cut his skin, not too deep, but just enough. He told them to collect the liquid that flowed from the cut. It would be clear as water and cold and just a little sweet. He told them to boil the liquid until it became a dark, thick, sweet syrup. They could eat this food, or they could boil it more until it became even thicker. If they poured the syrup into a trough and stirred it back and forth, it would turn into sugar.

The family did exactly as they were taught. They made maple syrup and sugar. They also made candy. Now they had enough food to keep them strong until the ice on the lake broke and there would be fish. The man tree had saved their lives.

Ininatig's gift taught the family to thank the trees each spring. From that day on, the family and all the others who learned about this food never forgot. Every spring the people hold a thanksgiving for the maple trees. Anishinabe elders teach that we depend on plants and animals. These gifts must not be taken for granted. Because people can forget, the thanksgiving story is told to each new generation of children.

A sugarbush is a place in the woods where sugar maple trees grow and people go year after year to make sugar. Sugar maples are known as hard maples. Their Latin name is *Acer saccarum*. The trees grow from eastern Canada across the northeastern part of the United States, as far south as Ohio and Virginia. Pollution from automobiles is killing many maple trees, however.

Sugar maples stand straight and tall, and their gray bark looks rough. In the summer, the seeds spin off the trees like tiny helicopters. The leaves of Ininatig look like open hands. In the fall, the leaves turn bright red, yellow, and orange.

The clear liquid that a maple tree produces in the spring is called *sap*. Sap is boiled to make maple syrup and sugar. The syrup is sweet and thick and tastes delicious on pancakes.

Porky White

It might seem easy to cut the trees, boil the sap, and make the sugar. But it is difficult to know which trees to cut, how long to boil the sap, and when to stir the syrup so it becomes sugar. Some people do know all these things. One of them is Gahgoonse, or Little Porcupine. His friends call him Porky. Others know him as Walter White.

Porky grew up in northern Minnesota in a place so famous for its maple sugar that it's called "Sugar Point." There, on the shores of Leech Lake, the land is just right for maple trees to grow. Porky learned about maple trees and sugaring by watching carefully and listening to his elders as they worked in the sugarbush.

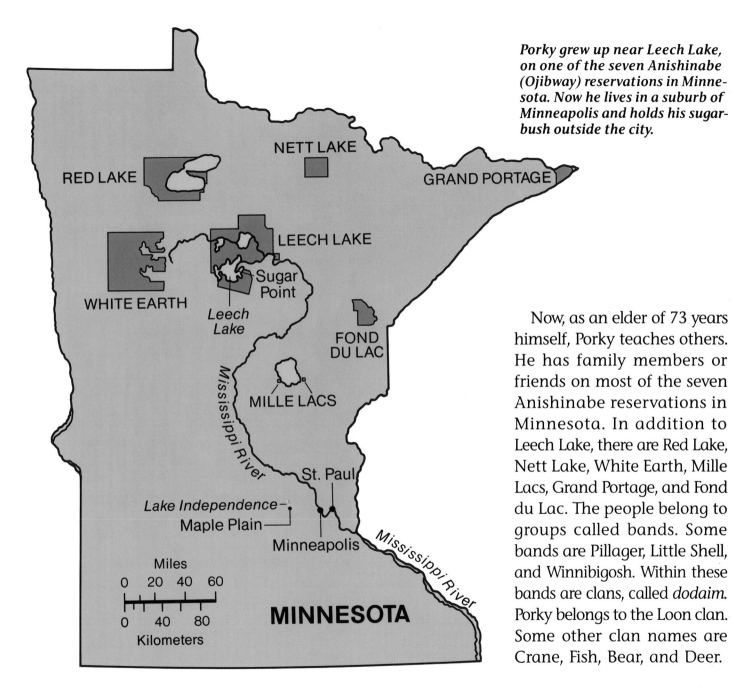

Porky grew up near Leech Lake, on one of the seven Anishinabe (Ojibway) reservations in Minnesota. Now he lives in a suburb of Minneapolis and holds his sugarbush outside the city.

NETT LAKE

RED LAKE

GRAND PORTAGE

LEECH LAKE

Sugar Point

WHITE EARTH

Leech Lake

FOND DU LAC

Mississippi River

MILLE LACS

St. Paul

Lake Independence
Maple Plain

Minneapolis

Mississippi River

Miles
0 20 40 60

0   40   80
Kilometers

MINNESOTA

Now, as an elder of 73 years himself, Porky teaches others. He has family members or friends on most of the seven Anishinabe reservations in Minnesota. In addition to Leech Lake, there are Red Lake, Nett Lake, White Earth, Mille Lacs, Grand Portage, and Fond du Lac. The people belong to groups called bands. Some bands are Pillager, Little Shell, and Winnibigosh. Within these bands are clans, called *dodaim*. Porky belongs to the Loon clan. Some other clan names are Crane, Fish, Bear, and Deer.

15

Left: *Usually when a sugarbush takes place the ice hasn't melted on the lake.* Below: *Students of all ages come to the camp to work— and have fun.*

Each spring Porky holds his own sugarbush camp, by Lake Independence, in Maple Plain, Minnesota. Hundreds of children from schools in the Twin Cities of Minneapolis and St. Paul come to the camp to learn about sugarmaking. Children and adults alike enjoy visiting the camp. But it is a working camp. The serious business of preparing the sugar requires the work of ten or more children and adults. Some people spend two days at the camp. Others live there for the entire two or three weeks of sugarmaking.

Porky and a friend cut a hole in the barrel in which sap will be boiled.

*B*efore the sugarbush camp starts in the spring, a lot of work has already been done. Porky plans his sugarbush in the summer, when the leaves are still on the trees. He checks to see if there are enough metal taps and plastic bags for collecting the sap. He inspects the storage barrels and the boiling barrels. The storage barrels are made of heavy plastic, and the boiling barrels are large steel drums turned on their sides.

Then, just after the winter holidays, Porky and his friends study the sugarbush catalog that has arrived in the mail. They make a list of what they want to order: more bags and taps, filters to strain the hot syrup, and dozens of quart and pint containers to pour the syrup into as it comes off the fire.

The wooden trough is cleaned up.

There are other things to make, mend, or buy. Porky's nephew Jim, a carver, checks the big wooden trough to see if it needs repair. He decides to make a new stirring paddle for testing the thickness of the boiling sap. Jim inspects the strainer to see if it will work for another season. He also makes several dozen small cones from fresh birchbark. Each cone will hold an ounce or two of maple candy.

Jim sends the order off to the sugarbush supply company in Wisconsin, and Porky watches the weather. As spring approaches, he takes a trip out to Lake Independence and looks for clues about whether the time is right to cut the trees. He sees big black crows on the lake and a woodpecker hunting for insects. The ice on the lake looks black. These clues are very important. The ice darkens as it starts to melt. The frozen ground is also beginning to thaw.

As the daytime weather grows sunny and warm, the maple trees begin to pump sap up from their roots into their trunks. Sap is the food that helps the trees make their leaves. During the winter, the sap is stored in the roots. In the spring, when the sap flows into the trunk and branches, it wakes the sleeping insects who live inside the tree. As the insects move around, woodpeckers can hear them. The birds tap at the tree to make the insects come out.

When Porky notices all these signs, he knows the trees are ready to tap. Tapping is the word for cutting the tree. A small hole is drilled in the tree, and a metal tap is pounded into the hole. The tap looks like a small faucet. As soon as the tree is tapped, the cold sap begins to drip from it.

The trees might be ready to tap in early or late March or even April. One thing is sure—you can't decide when to make a sugarbush just by looking at the calendar. You must be skilled like Porky and able to notice all the clues.

The maple trees, bare of their leaves, creak as they sway in the wind. The air smells fresh and it is still very cold, but the sun is shining. Everything seems right to begin this year's sugarbush.

Sap drips from the metal tap.

People begin to arrive at the camp, bringing necessary equipment, such as a log splitter.

Porky and Jim are the first to arrive at the sugarbush. Then Porky's other nephews Glen and Terrance White come, bringing their log-splitting equipment. Madeline Moose, Porky's sugaring companion of many years, arrives with kettles, dishes, pans, and the first week's supply of food. Other family members and friends come, too. Some drive down from Leech Lake, 200 miles north. Others come from Minneapolis, the largest city in Minnesota. Porky drives in from his home in Rosemount, a suburb of Minneapolis 45 miles from Maple Plain.

21

To tap the tree, Porky first drills a hole into it.

Before any supplies are unpacked, Porky taps one of the trees. The sap is running freely. It is time to thank the trees. The small group of people stand close to each other, their feet shifting on the cold ground. Porky speaks the Ojibway words. There is no other sound in the woods. It is as if everything has stopped to hear Porky give thanks to Ininatig— the tree that gives life to the people. Then, with a big smile, Porky announces that the sugarbush is about to begin!

Next, Porky pounds a metal tap into the hole he drilled. Then he hangs a clear plastic bag from the tap, so the sap can drip into the bag.

Terrance builds a fire to thaw the frozen ground so that pits can be dug for the cooking and boiling areas. Glen makes *nibish*—tea—from the first of the new sap that has been collected. He boils the sap in a big blue coffeepot and adds tea leaves. Everyone drinks the hot, sweet tea and eats some sandwiches that Madeline has made.

Suddenly, the roar of a chain saw breaks the peaceful silence. Porky has started cutting logs from the dead trees that fell during the winter. The camp will need a huge stockpile of wood to keep the fire burning for at least two weeks. In good years, when the winter has been very cold and the snow deep, the sugarbush goes on for three weeks.

Porky takes a break before he begins to chop wood.

Porky and a student saw and split logs for the fire.

The tipi is set up. First, lodge poles are arranged in a cone shape and tied together at the top. Then a strong rope is draped over the sticks and secured to the ground.

Glen, Jim, and Terrance put up the tents and the tipi. The food tent is packed with supplies for the days of hard work ahead. The first evening is full of happy talk: "This will be a good year," Jim says. "The sap is really running today," says Madeline. "Tomorrow will be even better."

At night, when Porky crawls into his sleeping bag, he hears the wind blow the branches of the trees, making a creaking sound like a wooden ship tossing on the waves. The sound is comforting, and soon Porky falls asleep.

*T*he next morning, Porky tells Jim that he dreamed about his boyhood, when his whole family camped at the sugarbush for weeks and weeks. What fun that was! Instead of buying plastic bags and barrels, Porky's mother made birchbark baskets called *muk kuks*. The muk kuks looked a lot like shoeboxes, with handles at the sides. They were pretty—dark orange inside with gray birchbark outside. Porky's family used muk kuks to catch the sap as it dripped from the trees.

Jenny, Porky's mother, also made bucket-shaped baskets for storing the finished maple sugar. Cut-out rounds of birchbark were sewn on as lids. She also rolled birchbark into small cones to be filled with maple sugar candy.

To make one muk kuk of finished sugar, Jenny needed to boil about 40 muk kuks of sap. When the sap boiled, water evaporated from it, leaving the sweet syrup. More boiling left a thick, gooey syrup that she poured into a wooden trough and stirred until it became sugar.

It was Porky's job, along with his brothers, to gather the piles and piles of wood needed for the fires. He also watched the fire and the boiling sap. Because the sap flowed only for a limited time, the work had to be done without stopping. Most of the time the family worked all day and into the night.

One day, Porky would hear the sound of peeping frogs. This meant the sap would soon stop flowing and the leaves of the man tree were ready to unfold from their buds. The warmest part of spring would soon come, and the sugarbush would be over for another year.

Porky and his family worked year after year at Sugar Point. Year after year they made sugar for themselves and their friends. A family of five like Porky's needed about 20 muk kuks of sugar for the year. After each sugarbush, the family held a feast that always included lots of fresh sugar.

In Porky's dream, he remembered the dark sweetness of the syrup. He thought it held the smell of the open fire and the memories of how much fun he had had.

Fresh maple sugar is a golden brown color.

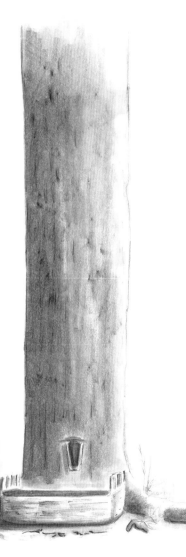

Sap drips through the metal tap into a plastic bag. In the past, Anishinabe people collected sap by inserting a smooth piece of wood into a cut in the tree. The sap ran from the wood into muk kuks.

The next morning at the camp is bright and sunny, just as everyone hoped it would be. The sap glistens in the clear plastic bags hanging from the taps. The bags high on the ridge fill up fast as the warm sun heats up the trees and the sap moves into the trunks. Further down the hill, in the shade, the sap drips more slowly, drop by drop. The ideal weather for a sugarbush is warm days and cold nights. That's what the trees like best.

Left: *Even though everyone helps out at the camp, there's still time for a game.*
Below: *A student tends the fire.*

Soon, the first busload of students arrive from the city. Porky teaches them the rules of the camp: no littering, everyone helps with the sugarmaking, and older students help younger ones—the bags of sap are heavy.

The trees must be treated with care. Living trees are never cut down. Only dead trees are used for firewood. Porky teaches that respecting Ininatig means keeping the woods clean. By treating Ininatig well, the people can continue to receive the gifts of *ziwagamizigun*—maple syrup—and *zizibahquat*—maple sugar.

The first thing the campers learn is how to tap the trees. The hand drill is twisted into the trunk, sending bits of wood flying. Then the metal tap is pounded in, and the sap begins to drip from the spout. Small trees are not tapped. They need all their sap for their growth. Bigger trees are strong enough to be tapped. In the past, a tree was cut close to the ground, a flat piece of wood was placed in the cut, and the sap drained into muk kuks.

But today, the students are hanging plastic bags from metal taps to collect the precious sap. As the bags fill up, they are emptied into buckets. The buckets are in turn emptied into larger barrels where the sap is held for boiling.

The sap boils all day and night. Glen and Terrance pour heavy loads of fresh, cold sap into the sap that is already boiling. Slowly the water in the sap evaporates as steam and the sweet smell of maple fills the air. Now the boiling liquid quickly becomes darker brown and thickens. Big bubbles rise to the surface of the barrel.

It takes a long time to boil all the sap.

Left: *The syrup is strained through cloth to remove anything floating in it.* Below: *The strained syrup is poured into jars.*

36

Porky watches the process very closely. Only he knows for sure when to take the boiling syrup off the fire. He does not use special tools or measurements. He knows from experience exactly how the syrup should look and feel. When Porky gives the signal, the syrup is taken off the fire. It is strained to remove any flecks or pieces of bark. The workers slowly pour the hot syrup into bottles and screw the lids on tightly.

The taste of fresh syrup— delicious!

As the syrup continues to boil, it becomes lighter in color and bubbles up. At the right moment, it is removed from the fire and poured into the wooden trough.

Sugarmaking is even more difficult than making syrup. To make sugar, Porky pours the syrup from the barrel into a smaller kettle. The syrup continues to boil—but it must not burn. As the bubbles get smaller and froth rises in the pot, Porky suddenly orders everyone to stand back. He removes the kettle from the fire and hands it to Jim, who pours the syrup into a long wooden trough. Workers stir the thick, golden syrup back and forth. Like magic, the liquid gradually changes into pure maple sugar. With more stirring, the lumps are smoothed out of the golden grains. Now the sugar is cool and dry. No one can resist eating a pinch.

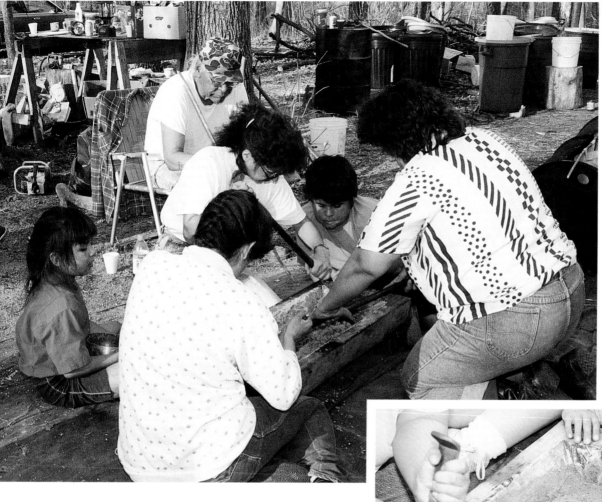

The workers stir the syrup until it turns into sugar.

Between the syrup and sugar stages, the syrup is the right consistency to make candy. Glen pours the candy into molds the shape of maple leaves.

Everyone wants a chance to lick the candy bowl.

As the syrup is boiling down for the sugarmaking, it passes through an in-between stage. It is not as thin as pancake syrup, but not yet thick enough to make sugar. This stage is the time to make candy. If there's snow on the ground, the syrup can be poured right onto it. It cools into squiggly shapes that taste good. Some of the syrup is poured into birchbark cones. The dark brown maple candy is tasty warm or cold.

The morning passes quickly. By noon everyone is hungry. Madeline Moose has prepared lunch for the workers. There are mugs of warm chicken stew and hot maple sap tea. Crispy apples are passed out for dessert. Now it is time for a story.

Porky settles the children down near the fire and tells them to listen carefully. Everyone is full of lunch and a little sleepy, but curious to know what the story is about. Porky begins.

There was a little boy who was the youngest in his family. His big brothers always got the best of him because they were bigger and stronger. The boy couldn't wait to be grown-up so he wouldn't get picked on so much.

One spring, an elder woman who lived nearby asked the boy's parents if he could be loaned to her to help with her sugarbush. She said she was getting too weak to do all the heavy work. The family agreed, and the boy was sent to her.

At first he was shy. He didn't know this woman very well. But he wanted his parents to be proud of him, so he tried to act very grown-up. The woman was surprised by this little boy who tried to behave like a man.

The sugarbush season went well, and there was a lot of sugar that year. The boy returned to his family full of stories about all the work he had done for the woman. His parents were very pleased.

Weeks went by, then one day a package came in the mail. It was a big package, and it was for the little boy. Inside was a warm coat—but it was much too big for the little boy. His face sank in disappointment.

Then his mother pulled out a card that was packed in the box with the coat. The card said, "Megwitch for loaning me your son. He is much bigger on the inside than he is on the outside, but one day soon, this coat will fit him. When that day comes, the outside will catch up with the inside."

Porky smiles and points to the sky. It is already late afternoon. These children have to get on the bus and go back to the city. Their sugarbush adventure is over, so soon.

The days of the sugarbush come and go. More children arrive to help. Some visit the camp with their teachers, others with their families. They all help and learn at the sugarbush, and each day the work goes on.

Sugarmaking is so much fun that everyone wishes it would never end. The fire crackles and the woods smell fresh and cold. School and home seem far away. Every night, the creaking trees put the workers to sleep, except those who stay up to take turns tending the fire.

But at last the sap stops flowing. Porky gives gifts of syrup and sugar to his many helpers. On the last day of the sugarbush, the people hold a feast of thanksgiving at the camp. There in the woods, with the sun shining and the birds singing, the workers and their families and friends gather with Porky to thank the trees once again. The people have taken good care of the maple trees this year.

Ininatig will be there again for them next year.

Word List

Acer saccarum—Latin name for the sugar maple tree

Anishinabe (ah-nish-ih-NAH-bay)—American Indian people from the northern Great Lakes area; sometimes called Ojibway or Chippewa

dodaim (doh-DAME)—the Ojibway word for "clan," a small, family-centered group of people

Ininatig (e-nee-NAH-tig)—the Ojibway word for the sugar maple tree; "inina" means "man," and "tig" means tree

megwitch (ME-gwitch)—the Ojibway word for "thank you"

muk kuk—the Ojibway word for "basket," especially the deep basket used to hold sugar

Ojibway—the language of the Anishinabe people

nibish (nee-BISH)—Ojibway word for tea

reservations—areas of land that Indian people kept through treaties with the United States government

sap—water that contains sugar and minerals and provides the food for maple trees; it is used to make maple syrup and sugar. Many trees, including other types of maples, produce sap, but only the sap of the sugar maple is sweet.

sugarbush—a woods where maple trees grow and maple sugar is made

tap—to drill into a sugar maple tree to draw off some of the sap

ziwagamizigun (zee-wahga-mee-zee-GUN)—Ojibway word for maple syrup

zizibahquat (zee-zee-BAH-quat)—Ojibway word for sugar

For Further Reading

Burns, Diane L. *Sugaring Season: Making Maple Syrup.*
    Minneapolis: Carolrhoda Books, 1990.

Johnston, Basil H. *Tales the Elders Told: Ojibway Legends.*
    Ontario: Royal Ontario Museum, 1981.

Osinski, Alice K. *The Chippewa.* Chicago: Childrens Press, 1987.

Tanner, Helen Hornbeck. Frank W. Porter III, ed. *The Ojibwa.*
    New York: Chelsea House, 1992.

We Are Still Here: Native Americans Today

Children of Clay
A Family of Pueblo Potters
    Rina Swentzell
    photographs by Bill Steen

Clambake
A Wampanoag Tradition
    Russell M. Peters
    photographs by John Madama

Ininatig's Gift of Sugar
Traditional Native Sugarmaking
    Laura Waterman Wittstock
    photographs by Dale Kakkak

Kinaaldá
A Navajo Girl Grows Up
    text and photographs by
    Monty Roessel

The Sacred Harvest
Ojibway Wild Rice Gathering
    Gordon Regguinti
    photographs by Dale Kakkak

Shannon
An Ojibway Dancer
    Sandra King
    photographs by Catherine Whipple

About the Contributors

Laura Waterman Wittstock is a founder of MIGIZI Communications in Minneapolis, where she serves as president and executive producer of radio programming. She writes a monthly column for *The Alley,* a Minneapolis neighborhood newspaper, and she edits the *Communicator* newsletter, published by MIGIZI. Wittstock is an enrolled member of the Seneca Nation in New York. Her father's family was part of the Stockbridge-Munsee relocation, and her mother is a descendant of Cornplanter. In addition to articles and chapters on native people, Ms. Wittstock writes on women, politics, and public policy.

Photographer **Dale Kakkak** was born and raised in the Menominee Nation. He is the staff photographer for *The Circle,* a native newspaper in the Twin Cities. He is a poet, fiction writer, and student of life.

Series Editor **Gordon Regguinti** is a member of the Leech Lake Band of Ojibway. He was raised on Leech Lake Reservation by his mother and grandparents. His Ojibway heritage has remained a central focus of his professional life. A graduate of the University of Minnesota with a B.A. in Indian Studies, Regguinti has written about Native American issues for newspapers and school curricula. He served as editor of the Twin Cities native newspaper *The Circle* for two years and is currently executive director of the Native American Journalists Association. He lives in Minneapolis and has six children and one grandchild.

Series Consultant **Juanita G. Corbine Espinosa,** Dakota/Ojibway, is the director of Native Arts Circle, Minnesota's first statewide Native American arts agency. She is first and foremost a community organizer, active in a broad range of issues, many of which are related to the importance of art in community life. In addition, she is a board member of the Minneapolis American Indian Center and an advisory member of the Minnesota State Arts Board's Cultural Pluralism Task Force. She was one of the first people to receive the state's McKnight Human Service Award. She lives in Minneapolis.

Illustrator **Carly Bordeau** is a member of the Anishinabe Nation, White Earth, Minnesota. She is a freelance graphic designer, illustrator, and photographer and the owner of All Nite Design and Photography. Carly graduated from the College of Associated Arts in St. Paul with a B.A. in Communication Design. She lives in St. Paul.